Survey of US Higher Education Faculty 2025, Use of AI Prompt Libraries

ISBN 979-8-88517-310-0

Table of Contents

QUESTIONNAIRE

1. Please answer the following:

   A. Name:
   B. Institution
   C. Title (associate, professor, instructor, etc.)
   D. Academic Field
   E. Email Address:

2. Which Carnegie Class description best fits your college?

   A. Community College
   B. BA-Granting Institution
   C. Specialized College (i.e. performing arts, seminary, etc.)
   D. MA-PHD Granting College
   E. Level 1 or Level 2 Research University

3. Your college is:

   A. Public
   B. Private

4. What is the approximate full time equivalent enrollment of your college?

5. Which age band best describes you?

   A. 30 or Under
   B. 31-39
   C. 40-49
   D. 50-59
   E. 60 or over

6. Which phrase best describes your political views?

   A. Left of Liberal
   B. Liberal
   C. Middle of the Road
   D. Conservative
   E. Right of Conservative

7. Which phrase best describes your annual personal income from all sources? (Do not include income from spouse or other life partner)

A. Less than $50,000
B. $50,000 to $75,000
C. $75,000+ to $100,000
D. $100,000 to $150,000
E. More than $150,000

8. You are:

A. Tenured
B. Not tenured but on a tenure track
C. Not tenured and not on a tenure track

9. Which phrase best describes your gender?

A. Female
B. Male
C. Transgender

10. Which phrase best describes your race or ethnicity?

A. White or Caucasian
B. Black or African American
C. Hispanic or Latino
D. Asian or Asian American Indigenous
E. Mixed Race

11. Your academic field is best described by which of the following designations

A. Social Sciences (except economics and psychology)
B. Economics/Finance
C. Business (other than economics/finance)
D. Performing Arts
E. Psychology & Social Work
F. Medicine
G. Law
H. Mathematics/Statistics/Computer Science
I. Hard Sciences (Chemistry, Biology, Physics)
J. Education or Library Science
K. Allied Health: Nursing, Occupation or Physical Therapy, Nutrition, etc. Earth and Environmental Sciences
L. Architecture, Fine and Visual Arts History
M. English, Literature & Languages
N. Other (please specify)

12. For how many years have you worked at your present institution?

A. 1 year or less
B. 1+ years to 3 years
C. 3+ years to 5 years
D. 5+ years to 10 years
E. 10+ years to 20 years
F. More than 20 years

13. Have you ever used a prompt library of any kind to obtain artificial intelligence prompts?

A. Yes
B. No

14. Have you ever purchased an artificial intelligence prompt?

A. Yes
B. No

15. Have you ever created an artificial intelligence prompt that was later saved in any kind of artificial intelligence prompt archive intended for use by others?

A. Yes
B. No

16. Have you ever shared or used a prompt through any of the following sources? (Check all that apply)

A. GitHub
B. Academic Library
C. Academic Department
D. Institutional Digital Repository
E. Instructional Design or Teaching & Learning Center
F. Professional or Trade Association
G. Other (Describe)

17. Have you ever included details of prompt engineering in any academic paper or article you have written?

A. Yes
B. No

18. How satisfied are you with the availability of libraries or archives of artificial intelligence prompts?

A. Very satisfied
B. Satisfied
C. Neither satisfied nor dissatisfied
D. Dissatisfied
E. Very dissatisfied

Characteristics of the sample

Age of respondent

	#
30 or under	9
31-39	40
40-49	103
50-59	155
60 or over	347
Unknown	6
Total	660

Type of college or Carnegie Class

	#
Community College	37
BA-Granting Institution	55
Specialized College (i.e. performing arts, seminary, etc.)	18
MA-PHD Granting College	133
Level 1 or Level 2 Research University	417
Total	660

Public or private college

	#
Public	454
Private	206
Total	660

Enrollment level of college

	#
Less than 5,000	168
5,000+ - 17,000	167
17,000+ - 32,000	168
More than 32,000	157
Total	660

Political views

	#
Left of Liberal	101
Liberal	255
Middle of the Road	193
Conservative	100
Unknown	11
Total	660

Annual personal income

	#
Less than $50,000	16
$50,000 to $75,000	65
$75,000+ to $100,000	116
$100,000 to $150,000	201
More than $150,000	247
Unknown	15
Total	660

Gender of respondent

	#
Female	272
Male	376
Transgender	5
Unknown	7
Total	660

Race or ethnicity

	#
White or Caucasian	536
Black or African American	26
Hispanic or Latino	27
Asian or Asian American	38
Mixed Race	18
Unknown	15
Total	660

Academic major

	#
Social Sciences (except economics and psychology)	65
Economics/Finance	18
Business (other than economics/finance)	65
Performing Arts	20
Psychology & Social Work	31
Medicine	78
Law	6
Mathematics/Statistics/Computer Science	35
Hard Sciences (Chemistry, Biology, Physics)	103
Education or Library Science	42
Allied Health: Nursing, Occupation or Physical Therapy, Nutrition, etc.	44
Earth and Environmental Sciences	22
Architecture, Fine and Visual Arts	10
History	14
English, Literature & Languages	49
Communications & Journalism	11
Engineering	37
Philosophy & Religion	6
Other	4
Total	660

Years at present institution

	#
3 years or less	40
3+ years to 5 years	28
5+ years to 10 years	77
10+ years to 20 years	201
More than 20 years	314
Total	660

INSTITUTIONAL AFFILIATIONS OF THE SURVEY PARTICIPANTS

A.T. Still University
A.T. Still University/KCOM
Albert Einstein College of Medicine
Alfred University
American University of Iraq-Baghdad
American University Washington College of Law
APUS
Arizona State U
Arkansas State University-Mountain Home
Ashland University
Austin Community College
Babson College
Banner UMC - University of Arizona
Baylor College of Medicine
Belmont University
Blando
Blinn College Small Business Development Center
Boston Children's Hospital
Bowie State University
Brigham Young University
Brooklyn College
California College of the Arts
California Community Colleges
California Lutheran University
California State University Channel Islands
California State University, Northridge, CA
Caltech
Carnegie Mellon University
Cedars-Sinai Health Sciences University
Cedars-Sinai Medical Center/ UCLA
Central Michigan University
Central New Mexico Community College
Central Texas College
Champlain College
Chapman university
Charleston Southern University
CIRES
City University of New York
Claborn
Clemson University
Cleveland State University
Colby College
College of William & Maru

Colorado School of Public Health
Columbia University
Cornell University
CSU East Bay
CU Boulder
CU Medicine
CU Sports Medicine & Performance Center
Cummings School of Veterinary Medicine at Tufts University
CUNY Graduate Center
CUNY Hunter College
Dallas Baptist University
Dartmouth College
DePaul University
Dodge City Community College
Dominican University
Dona Ana Community College/NMSU
Drexel University
Duke University School of Medicine
Eastern Illinois University
Edinboro University
Embry-Riddle Aeronautical University Prescott
Erie Community College
Fairmont State University
Faulkner University
Fayetteville State University
Florida Atlantic University
Florida International University
Florida State University
Fordham University
Fort Hays State University
Front Range Community College
Frontier Nursing University
Garry Schyman Productions
Gavilan College
Georgetown University
Georgia Tech
Grand Valley State University
Hampton University
Harper College
Harvard Medical School
Hawaii Pacific University
Hilbert College
Hillary
Hobart and William Smith Colleges
Hofstra University
Holy Family University

Hostos Community College, City University of New York
Howard University
Icahn School of Medicine
Illinois Institute of Technology
Illinois State University
Indiana University
IU Indianapolis
James Madison University
Jefferson
John Jay College
Johns Hopkins University
Juniata College
Kansas State University - Manhattan, KS
Kean University
Kennesaw State University
Kent State University
Keystone College
Kutztown University of PA
LaGuardia Community College (CUNY)
Lake Washington Institute of Technology
Lewis University
Lindenwood University
Louisiana State University
Louisiana Tech University
Loyola University Chicago
Loyola University Maryland
MA College of Pharmacy and Health Sciences
Macomb Community College
Manchester University
Marymount University
Mayville State University
McKendree University
Medical College of Wisconsin
Merritt College
Metropolitan College of New York
Michigan State University
Middle Georgia State University
Middle Tennessee State University
Midlands Technical College
Miller
Milligan University
Misericordia University
Mississippi Gulf Coast Community College
Mississippi State University
Missouri S&T
Missouri Southern State University

Missouri State University
Monmouth College
Montana State University
Montana Technological University
Montefiore Einstein
Mount Mary University
National Louis University
National University
New England College
New York Medical College
North Carolina A&T State University
North Central TX College
Northeastern University
Northern Arizona University
Northwestern University
Norwich University
Nova Southeastern University
Ohio University
Oklahoma State University-Oklahoma City
Old Dominion University
Oregon Health and Science University
Pace University
Penn State University
PennWest
Point Park University
Queens College, CUNY
Rollins College
Roosevelt University
Rose-Hulman Inst. of Tech
Rush University Medical Center
Saint Leo University
Sam Houston State University
San Francisco State University
Seattle Pacific University
Seton Hall University
Sinclair Community College
Smith College
Southern New Hampshire University
St. Mary's of Texas
State University of New York at Buffalo
SUNY at Buffalo
Syracuse University
Temple University
Texas A&M University
Texas Christian University
Texas Lutheran University

Texas Southern University
Texas Tech University
Texas Woman's University
The City College of New York
The CUNY Graduate Center
The Ohio State University Wexner Medical Center
The University of Alabama
The University of Arizona
The University of Findlay
The University of Texas at Tyler
Thomas More University
Tidewater Community College
Tiffin University
Tompkins Cortland community college
Touro University
Tufts University
UCBA
UCCS
UCLA MSS
UCSD
UMass Chan Medical School
UMass Lowell
Univ Alabama at Birmingham
Univ of Colorado Boulder
Univ. of Massachusetts Dartmouth
University at Buffalo
University of Arizona
University of Arkansas
University of Cincinnati
University of Colorado
University of Connecticut
University of Denver
University of Florida
University of Georgia
University of Houston-Clear Lake
University of Illinois
University of Iowa
University of Maryland Global Campus
University of Massachusetts Boston
University of Mississippi
University of Missouri
University of Nebraska - Lincoln
University of Nebraska at Omaha
University of Nevada Reno
University of New Hampshire
University of New Mexico

University of North Carolina at Chapel Hill
University of Oregon
University of Pennsylvania
University of Phoenix
University of Pittsburgh
University of Rochester
University of Rochester Medical Center
University of Sothern California
University of South Carolina
University of South Dakota
University of South Florida
University of Tennessee
University of Texas at Arlington
University of Texas at Austin
University of Texas at Tyler
University of the Incarnate Word
University of Toledo
University of Virginia
University of Washington
University of Wisconsin - Milwaukee
USC - School of Cinematic Arts
USC Mann
UT Dallas
Utah Valley University
Valencia College
Vanderbilt University Medical Center
Villanova University
Virginia Tech
Wake Forest School of Medicine
Wartburg College
Washington University
Wayne State University
West Virginia University
Western Carolina University
Western Seminary
Westfield State University
Wheeling University
Wichita State University
Williams
Yale School of Public Health
Zucker School of Medicine
Zucker School of Medicine/Hofstra

SUMMARY OF MAIN FINDINGS

Section 1: Use of Prompt Libraries, Summary of Survey Findings

Q1. Have you ever used a prompt library of any kind to obtain AI prompts?

- Overall: Only 6.7% said *Yes*, 89.9% said *No* (3.5% no answer).
- By Institution Size: Usage highest in mid-range colleges (5K–17K enrollment at 9.0% Yes), and lowest in the largest institutions (over 32K at 4.5% Yes).
- By Carnegie Type: BA-granting institutions lead (12.7% Yes); research universities lower (5.0% Yes).
- Public vs Private: More usage reported at private institutions (8.3%) than public (6.0%).
- By Age: Faculty aged 40–59 show slightly elevated usage (~6.8–7.7%).
- By Politics: Conservatives report highest usage (8.0% Yes), while left/liberal respondents report lower (~4–7%).
- By Income: Highest usage among respondents earning <$75K/year (~12.3%) but declines at higher income levels.
- Tenure Status: Non-tenure-track respondents report higher usage (9.1% Yes).
- Gender: Slightly more use among female faculty (7.2%) than male (6.3%).
- Ethnicity: Highest usage among Black respondents (11.5%), and Asian (7.9%).
- Academic Field: Top usage in Psychology & Social Work (16.1%), Business (10.8%), Education/Library Sci (11.9%), Fine & Visual Arts (10%), English/Languages (10.2%), Other fields (25%).
- Years at Institution: Slight increase with tenure at institution; >20 years shows 7.3% Yes.

Section 2: Purchasing AI Prompts, Summary of Survey Findings

Q2. Have you ever purchased an AI prompt?

- Overall: Very rare – only 1.8% Yes, 94.9% No (3.3% no answer).
- By Institution Size: Usage remains uniformly low (1.2%–2.4% across sizes).
- Carnegie Type: Slightly higher at MA-PhD colleges (2.3%) vs research universities (1.9%).
- Public vs Private: Similar (~1.8% Yes).
- Age: Peak at 40–49 (4.9%), but minimal in younger (0%) and older categories.
- Politics & Income: Modest variation (~0.9%–2.3% by ideology and income), with occasional outliers.
- Tenure Track: Negligible; non-tenure-track (3.0%) slightly higher.
- Demographics: Higher purchase rates among Black faculty (7.7%), Asian respondents (5.3%), especially in Allied Health (9.1%); rare otherwise.

Section 3: Creating and Sharing Prompts in Public Archives, Summary of Survey Findings

Q3. Have you ever created an AI prompt that was later saved in an archive for others?

- Overall: Around 5.2% Yes, 91.1% No.
- By Age: Highest in 40–49 range (13.6% Yes); younger and older very low.
- Politics: Left-of-liberal respondents slightly higher (7.9% Yes).
- Tenure Status: Both tenure-track and non-track hover ~8%.
- Race/Ethnicity: Black faculty (11.5%) and Asian (~5.3%) again standout.
- Fields: Strongest in Education/Library Science (11.9%), English/Languages (12.2%), Psych/Social Work (6.5%), and "Other" (25%).
- Years at Institution: Highest among mid-range (3+ to 5 years: 14.3% Yes).

Section 4: Channels for Sharing or Using Prompts, Summary of Survey Findings

(Respondents checked multiple sources; percentages represent share who have used each channel.)

- GitHub: Used by 2.4% overall. Higher in Education & Earth Sciences; lowest among policies.
- Academic Library: ~3.3% overall; strongest in Ed/Library Science, mid-income, public institutions.
- Academic Department: ~5.8% overall; fairly consistent across institution types and fields; higher among tenured and Black faculty.
- Institutional Digital Repository: Very low (~1.5% overall); modestly more in Ed/Library Science.
- Instructional Design Center: 4.1% overall; peaked in Ed/Library Sci (11.9%) and Allied Health.
- Professional or Trade Association: 3.6% overall; highest usage in Communications, Ed/Library Science (14.3%), Business (7.7%).
- Other (described): 4.1% overall; peaks in Engineering (8.1%), Medicine (9.0%), Business (7.7%).

Section 5: Academic Publication of Prompt Engineering Details, Summary of Survey Findings

Q5. Have you ever included prompt engineering details in an academic paper/article?

- Overall: Only 2.7% Yes, 91.7% No, with 5.6% no answer.
- Fields: Education/Library Science (11.9%), Psychology & Social Work (6.5%), Math/CS (2.9%), Business (4.6%). Very rare in STEM, Engineering, or Humanities.

- Demographics: Higher reporting among those with incomes <$75K, also among Asian respondents (5.3%), Black (7.7%), and tenured faculty.

Section 6: Satisfaction with Prompt Library Availability, Summary of Survey Findings

Q6. How satisfied are you with availability of prompt libraries?

- Overall: Vast majority neutral: 77.0% neither satisfied nor dissatisfied; only about 8% positive (1.5% very satisfied, 6.4% satisfied); ~5.5% dissatisfied.
- By Institution Size & Type: Minimal variation; slightly more dissatisfaction at largest institutions.
- Fields: Education/Library Science more positive (~14% satisfied/very satisfied); Philosophy & Religion least satisfied (16.7% very dissatisfied).
- Demographics: Conservative respondents more likely to be "very satisfied" (4%) than others. Black respondents mostly neutral (88%)—but only 11.5% satisfied. Asian faculty had high "no answer" (23.7%) and neutrality.
- Tenure Status: Non-tenure-track noted higher dissatisfaction (10.8%), while tenured respondents slightly more positive.
- By Years at Institution: Most neutral; those with 3+ to 5 years express the highest neutrality (~93%).

Key Findings

- Prompt-library use is rare: fewer than 7% have ever used one, and prompt purchases or publication of prompt techniques are even less common.
- Mid-career faculty and certain fields (Education, Psych/social work, English, Business) show relatively higher engagement in using, creating, or sharing prompts.
- Academic sharing is limited: GitHub and departmental exchange are the most common channels but still under 6% usage.
- Prompt engineering is rarely documented in formal research outputs.
- Most faculty feel largely indifferent about the current availability of prompt archives, indicating a perceived lack of both quality and visibility.

Table 1.1 Have you ever used a prompt library of any kind to obtain artificial intelligence prompts?

	No Answer	Yes	No
Entire sample	3.48%	6.67%	89.85%

Table 1.2 Have you ever used a prompt library of any kind to obtain artificial intelligence prompts? Broken out by enrollment

Enrollment	No Answer	Yes	No
Less than 5,000	2.38%	6.55%	91.07%
5,000+ - 17,000	2.40%	8.98%	88.62%
17,000+ - 32,000	3.57%	6.55%	89.88%
More than 32,000	5.73%	4.46%	89.81%

Table 1.3 Have you ever used a prompt library of any kind to obtain artificial intelligence prompts? Broken out by type of college or Carnegie Class

Type of college or Carnegie Class	No Answer	Yes	No
Community College	0.00%	5.41%	94.59%
BA-Granting Institution	0.00%	12.73%	87.27%
Specialized College (i.e. performing arts, seminary, etc.)	0.00%	5.56%	94.44%
MA-PHD Granting College	3.76%	9.77%	86.47%
Level 1 or Level 2 Research University	4.32%	5.04%	90.65%

Table 1.4 Have you ever used a prompt library of any kind to obtain artificial intelligence prompts? Broken out by public or private college

Public or private college	No Answer	Yes	No
Public	4.41%	5.95%	89.65%
Private	1.46%	8.25%	90.29%

Table 1.5 Have you ever used a prompt library of any kind to obtain artificial intelligence prompts? Broken out by age of respondent

age of respondent	No Answer	Yes	No
30 or under	0.00%	0.00%	100.00%
31-39	2.50%	2.50%	95.00%
40-49	2.91%	6.80%	90.29%
50-59	4.52%	7.74%	87.74%
60 or over	3.46%	6.92%	89.63%
Unknown	0.00%	0.00%	100.00%

Table 1.6 Have you ever used a prompt library of any kind to obtain artificial intelligence prompts? Broken out by political views

Political views	No Answer	Yes	No
Left of Liberal	2.97%	3.96%	93.07%
Liberal	3.53%	7.06%	89.41%
Middle of the Road	3.63%	6.74%	89.64%
Conservative	4.00%	8.00%	88.00%
Unknown	0.00%	9.09%	90.91%

Table 1.7 Have you ever used a prompt library of any kind to obtain artificial intelligence prompts? Broken out by annual personal income

annual personal income	No Answer	Yes	No
Less than $50,000	0.00%	12.50%	87.50%
$50,000 to $75,000	0.00%	12.31%	87.69%
$75,000+ to $100,000	3.45%	4.31%	92.24%
$100,000 to $150,000	4.48%	8.46%	87.06%
More than $150,000	3.24%	4.86%	91.90%
Unknown	13.33%	0.00%	86.67%

Table 1.8 Have you ever used a prompt library of any kind to obtain artificial intelligence prompts? Broken out by tenure track

Tenure track	No Answer	Yes	No
Tenured	3.87%	5.41%	90.72%
Not tenured but on a tenure track	2.70%	5.41%	91.89%
Not tenured and not on a tenure track	2.60%	9.09%	88.31%
Unknown	25.00%	0.00%	75.00%

Table 1.9 Have you ever used a prompt library of any kind to obtain artificial intelligence prompts? Broken out by gender of respondent

Gender of respondent	No Answer	Yes	No
Male	2.94%	6.25%	90.81%
Female	3.99%	7.18%	88.83%
Transgender	0.00%	0.00%	100.00%
Unknown	0.00%	0.00%	100.00%

Table 1.10 Have you ever used a prompt library of any kind to obtain artificial intelligence prompts? Broken out by race or ethnicity

Race or ethnicity	No Answer	Yes	No
White or Caucasian	3.54%	6.34%	90.11%
Black or African American	0.00%	11.54%	88.46%
Hispanic or Latino	0.00%	7.41%	92.59%
Asian or Asian American	10.53%	7.89%	81.58%
Mixed Race	0.00%	5.56%	94.44%
Unknown	0.00%	6.67%	93.33%

Table 1.11 Have you ever used a prompt library of any kind to obtain artificial intelligence prompts? Broken out by academic field

academic field	No Answer	Yes	No
Social Sciences (except economics and psychology)	1.54%	4.62%	93.85%
Education or Library Science	2.38%	11.90%	85.71%
Allied Health: Nursing, Occupation or Physical Therapy, Nutrition, etc.	4.55%	0.00%	95.45%
Earth and Environmental Sciences	4.55%	4.55%	90.91%
Architecture, Fine and Visual Arts	0.00%	10.00%	90.00%
History	0.00%	0.00%	100.00%
English, Literature & Languages	2.04%	10.20%	87.76%
Communications & Journalism	0.00%	9.09%	90.91%
Engineering	10.81%	5.41%	83.78%
Philosophy & Religion	0.00%	0.00%	100.00%
Economics/Finance	0.00%	5.56%	94.44%
Business (other than economics/finance)	1.54%	10.77%	87.69%
Performing Arts	0.00%	10.00%	90.00%
Psychology & Social Work	9.68%	16.13%	74.19%
Medicine	5.13%	2.56%	92.31%
Law	0.00%	0.00%	100.00%
Mathematics/Statistics/Computer Science	5.71%	5.71%	88.57%
Hard Sciences (Chemistry, Biology, Physics)	2.91%	5.83%	91.26%
Other	0.00%	25.00%	75.00%

Table 1.12 Have you ever used a prompt library of any kind to obtain artificial intelligence prompts? Broken out by years at present institution

Years at present institution	No Answer	Yes	No
3 years or less	5.00%	5.00%	90.00%
3+ years to 5 years	3.57%	3.57%	92.86%
5+ years to 10 years	0.00%	6.49%	93.51%
10+ years to 20 years	2.99%	6.47%	90.55%
More than 20 years	4.46%	7.32%	88.22%

Table 2.1 Have you ever purchased an artificial intelligence prompt?

	No Answer	Yes	No
Entire sample	3.33%	1.82%	94.85%

Table 2.2 Have you ever purchased an artificial intelligence prompt? Broken out by enrollment

Enrollment	No Answer	Yes	No
Less than 5,000	2.38%	1.79%	95.83%
5,000+ - 17,000	2.40%	1.20%	96.41%
17,000+ - 32,000	3.57%	2.38%	94.05%
More than 32,000	5.10%	1.91%	92.99%

Table 2.3 Have you ever purchased an artificial intelligence prompt? Broken out by type of college or Carnegie Class

Type of college or Carnegie Class	No Answer	Yes	No
Community College	0.00%	0.00%	100.00%
BA-Granting Institution	0.00%	0.00%	100.00%
Specialized College (i.e. performing arts, seminary, etc.)	0.00%	5.56%	94.44%
MA-PHD Granting College	3.76%	2.26%	93.98%
Level 1 or Level 2 Research University	4.08%	1.92%	94.00%

Table 2.4 Have you ever purchased an artificial intelligence prompt? Broken out by public or private college

Public or private college	No Answer	Yes	No
Public	3.96%	1.76%	94.27%
Private	1.94%	1.94%	96.12%

Table 2.5 Have you ever purchased an artificial intelligence prompt? Broken out by age of respondent

age of respondent	No Answer	Yes	No
30 or under	11.11%	0.00%	88.89%
31-39	2.50%	0.00%	97.50%
40-49	2.91%	4.85%	92.23%
50-59	4.52%	2.58%	92.90%
60 or over	2.88%	0.86%	96.25%
Unknown	0.00%	0.00%	100.00%

Table 2.6 Have you ever purchased an artificial intelligence prompt? Broken out by political views

Political views	No Answer	Yes	No
Left of Liberal	1.98%	0.99%	97.03%
Liberal	3.14%	2.35%	94.51%
Middle of the Road	3.63%	2.07%	94.30%
Conservative	5.00%	1.00%	94.00%
Unknown	0.00%	0.00%	100.00%

Table 2.7 Have you ever purchased an artificial intelligence prompt? Broken out by annual personal income

annual personal income	No Answer	Yes	No
Less than $50,000	0.00%	0.00%	100.00%
$50,000 to $75,000	0.00%	3.08%	96.92%
$75,000+ to $100,000	3.45%	0.86%	95.69%
$100,000 to $150,000	4.48%	2.99%	92.54%
More than $150,000	2.83%	0.81%	96.36%
Unknown	13.33%	6.67%	80.00%

Table 2.8 Have you ever purchased an artificial intelligence prompt? Broken out by tenure track

Tenure track	No Answer	Yes	No
Tenured	3.61%	1.03%	95.36%
Not tenured but on a tenure track	2.70%	2.70%	94.59%
Not tenured and not on a tenure track	2.60%	3.03%	94.37%
Unknown	25.00%	0.00%	75.00%

Table 2.9 Have you ever purchased an artificial intelligence prompt? Broken out by gender of respondent

Gender of respondent	No Answer	Yes	No
Male	2.57%	2.21%	95.22%
Female	3.99%	1.60%	94.41%
Transgender	0.00%	0.00%	100.00%
Unknown	0.00%	0.00%	100.00%

Table 2.10 Have you ever purchased an artificial intelligence prompt? Broken out by race or ethnicity

Race or ethnicity	No Answer	Yes	No
White or Caucasian	3.36%	1.49%	95.15%
Black or African American	0.00%	7.69%	92.31%
Hispanic or Latino	0.00%	0.00%	100.00%
Asian or Asian American	10.53%	5.26%	84.21%
Mixed Race	0.00%	0.00%	100.00%
Unknown	0.00%	0.00%	100.00%

Table 2.11 Have you ever purchased an artificial intelligence prompt? Broken out by academic field

academic field	No Answer	Yes	No
Social Sciences (except economics and psychology)	0.00%	3.08%	96.92%
Education or Library Science	2.38%	2.38%	95.24%
Allied Health: Nursing, Occupation or Physical Therapy, Nutrition, etc.	6.82%	9.09%	84.09%
Earth and Environmental Sciences	4.55%	4.55%	90.91%
Architecture, Fine and Visual Arts	0.00%	10.00%	90.00%
History	0.00%	0.00%	100.00%
English, Literature & Languages	2.04%	2.04%	95.92%
Communications & Journalism	0.00%	0.00%	100.00%
Engineering	10.81%	0.00%	89.19%
Philosophy & Religion	0.00%	0.00%	100.00%
Economics/Finance	0.00%	0.00%	100.00%
Business (other than economics/finance)	1.54%	0.00%	98.46%
Performing Arts	0.00%	0.00%	100.00%
Psychology & Social Work	6.45%	0.00%	93.55%
Medicine	5.13%	0.00%	94.87%
Law	0.00%	16.67%	83.33%
Mathematics/Statistics/Computer Science	5.71%	0.00%	94.29%
Hard Sciences (Chemistry, Biology, Physics)	2.91%	0.97%	96.12%
Other	0.00%	0.00%	100.00%

Table 2.12 Have you ever purchased an artificial intelligence prompt? Broken out by years at present institution

Years at present institution	No Answer	Yes	No
3 years or less	5.00%	0.00%	95.00%
3+ years to 5 years	3.57%	7.14%	89.29%
5+ years to 10 years	0.00%	2.60%	97.40%
10+ years to 20 years	2.99%	2.99%	94.03%
More than 20 years	4.14%	0.64%	95.22%

Table 3.1 Have you ever created an artificial intelligence prompt that was later saved in any kind of artificial intelligence prompt archive intended for use by others?

	No Answer	Yes	No
Entire sample	3.79%	5.15%	91.06%

Table 3.2 Have you ever created an artificial intelligence prompt that was later saved in any kind of artificial intelligence prompt archive intended for use by others? Broken out by enrollment

Enrollment	No Answer	Yes	No
Less than 5,000	3.57%	4.17%	92.26%
5,000+ - 17,000	2.99%	5.99%	91.02%
17,000+ - 32,000	4.17%	5.95%	89.88%
More than 32,000	4.46%	4.46%	91.08%

Table 3.3 Have you ever created an artificial intelligence prompt that was later saved in any kind of artificial intelligence prompt archive intended for use by others? Broken out by type of college or Carnegie Class

Type of college or Carnegie Class	No Answer	Yes	No
Community College	0.00%	2.70%	97.30%
BA-Granting Institution	1.82%	5.45%	92.73%
Specialized College (i.e. performing arts, seminary, etc.)	0.00%	5.56%	94.44%
MA-PHD Granting College	3.76%	6.02%	90.23%
Level 1 or Level 2 Research University	4.56%	5.04%	90.41%

Table 3.4 Have you ever created an artificial intelligence prompt that was later saved in any kind of artificial intelligence prompt archive intended for use by others? Broken out by public or private college

Public or private college	No Answer	Yes	No
Public	4.63%	5.29%	90.09%
Private	1.94%	4.85%	93.20%

Table 3.5 Have you ever created an artificial intelligence prompt that was later saved in any kind of artificial intelligence prompt archive intended for use by others? Broken out by age of respondent

age of respondent	No Answer	Yes	No
30 or under	0.00%	0.00%	100.00%
31-39	2.50%	2.50%	95.00%
40-49	3.88%	13.59%	82.52%
50-59	5.16%	7.74%	87.10%
60 or over	3.46%	2.02%	94.52%
Unknown	0.00%	0.00%	100.00%

Table 3.6 Have you ever created an artificial intelligence prompt that was later saved in any kind of artificial intelligence prompt archive intended for use by others? Broken out by political views

Political views	No Answer	Yes	No
Left of Liberal	1.98%	7.92%	90.10%
Liberal	4.31%	5.10%	90.59%
Middle of the Road	4.15%	4.66%	91.19%
Conservative	4.00%	4.00%	92.00%
Unknown	0.00%	0.00%	100.00%

Table 3.7 Have you ever created an artificial intelligence prompt that was later saved in any kind of artificial intelligence prompt archive intended for use by others? Broken out by annual personal income

annual personal income	No Answer	Yes	No
Less than $50,000	0.00%	0.00%	100.00%
$50,000 to $75,000	0.00%	7.69%	92.31%
$75,000+ to $100,000	3.45%	5.17%	91.38%
$100,000 to $150,000	4.98%	4.98%	90.05%
More than $150,000	3.64%	5.26%	91.09%
Unknown	13.33%	0.00%	86.67%

Table 3.8 Have you ever created an artificial intelligence prompt that was later saved in any kind of artificial intelligence prompt archive intended for use by others? Broken out by tenure track

Tenure track	No Answer	Yes	No
Tenured	4.38%	3.09%	92.53%
Not tenured but on a tenure track	2.70%	8.11%	89.19%
Not tenured and not on a tenure track	2.60%	8.23%	89.18%
Unknown	25.00%	0.00%	75.00%

Table 3.9 Have you ever created an artificial intelligence prompt that was later saved in any kind of artificial intelligence prompt archive intended for use by others? Broken out by gender of respondent

Gender of respondent	No Answer	Yes	No
Male	2.94%	5.15%	91.91%
Female	4.52%	5.05%	90.43%
Transgender	0.00%	20.00%	80.00%
Unknown	0.00%	0.00%	100.00%

Table 3.10 Have you ever created an artificial intelligence prompt that was later saved in any kind of artificial intelligence prompt archive intended for use by others? Broken out by race or ethnicity

Race or ethnicity	No Answer	Yes	No
White or Caucasian	3.92%	4.85%	91.23%
Black or African American	0.00%	11.54%	88.46%
Hispanic or Latino	0.00%	3.70%	96.30%
Asian or Asian American	10.53%	5.26%	84.21%
Mixed Race	0.00%	5.56%	94.44%
Unknown	0.00%	6.67%	93.33%

Table 3.11 Have you ever created an artificial intelligence prompt that was later saved in any kind of artificial intelligence prompt archive intended for use by others? Broken out by academic field

academic field	No Answer	Yes	No
Social Sciences (except economics and psychology)	3.08%	6.15%	90.77%
Education or Library Science	4.76%	11.90%	83.33%
Allied Health: Nursing, Occupation or Physical Therapy, Nutrition, etc.	4.55%	2.27%	93.18%
Earth and Environmental Sciences	4.55%	0.00%	95.45%
Architecture, Fine and Visual Arts	0.00%	0.00%	100.00%
History	0.00%	0.00%	100.00%
English, Literature & Languages	2.04%	12.24%	85.71%
Communications & Journalism	0.00%	0.00%	100.00%
Engineering	10.81%	5.41%	83.78%
Philosophy & Religion	0.00%	0.00%	100.00%
Economics/Finance	5.56%	0.00%	94.44%
Business (other than economics/finance)	1.54%	10.77%	87.69%
Performing Arts	0.00%	5.00%	95.00%
Psychology & Social Work	6.45%	6.45%	87.10%
Medicine	5.13%	3.85%	91.03%
Law	0.00%	0.00%	100.00%
Mathematics/Statistics/Computer Science	5.71%	0.00%	94.29%
Hard Sciences (Chemistry, Biology, Physics)	2.91%	1.94%	95.15%
Other	0.00%	25.00%	75.00%

Table 3.12 Have you ever created an artificial intelligence prompt that was later saved in any kind of artificial intelligence prompt archive intended for use by others? Broken out by years at present institution

Years at present institution	No Answer	Yes	No
3 years or less	5.00%	7.50%	87.50%
3+ years to 5 years	3.57%	14.29%	82.14%
5+ years to 10 years	2.60%	5.19%	92.21%
10+ years to 20 years	3.48%	7.46%	89.05%
More than 20 years	4.14%	2.55%	93.31%

Table 4 Have you ever shared or used a prompt through any of the following sources? (Check all that apply)

Table 4.1.1 Have you ever shared or used a prompt through any of the following sources? GitHub.

	No Answer	Yes	No
Entire sample	0.00%	2.42%	97.58%

Table 4.1.2 Have you ever shared or used a prompt through any of the following sources? GitHub. Broken out by enrollment

Enrollment	Yes	No
Less than 5,000	2.38%	97.62%
5,000+ - 17,000	1.20%	98.80%
17,000+ - 32,000	4.17%	95.83%
More than 32,000	1.91%	98.09%

Table 4.1.3 Have you ever shared or used a prompt through any of the following sources? GitHub. Broken out by type of college or Carnegie Class

Type of college or Carnegie Class	Yes	No
Community College	0.00%	100.00%
BA-Granting Institution	1.82%	98.18%
Specialized College (i.e. performing arts, seminary, etc.)	0.00%	100.00%
MA-PHD Granting College	2.26%	97.74%
Level 1 or Level 2 Research University	2.88%	97.12%

Table 4.1.4 Have you ever shared or used a prompt through any of the following sources? GitHub. Broken out by public or private college

Public or private college	Yes	No
Public	2.20%	97.80%
Private	2.91%	97.09%

Table 4.1.5 Have you ever shared or used a prompt through any of the following sources? GitHub. Broken out by age of respondent

age of respondent	Yes	No
30 or under	0.00%	100.00%
31-39	5.00%	95.00%
40-49	3.88%	96.12%
50-59	3.23%	96.77%
60 or over	1.44%	98.56%
Unknown	0.00%	100.00%

Table 4.1.6 Have you ever shared or used a prompt through any of the following sources? GitHub. Broken out by political views

Political views	Yes	No
Left of Liberal	0.00%	100.00%
Liberal	2.75%	97.25%
Middle of the Road	4.15%	95.85%
Conservative	0.00%	100.00%
Unknown	9.09%	90.91%

Table 4.1.7 Have you ever shared or used a prompt through any of the following sources? GitHub. Broken out by annual personal income

annual personal income	Yes	No
Less than $50,000	0.00%	100.00%
$50,000 to $75,000	6.15%	93.85%
$75,000+ to $100,000	3.45%	96.55%
$100,000 to $150,000	1.99%	98.01%
More than $150,000	1.62%	98.38%
Unknown	0.00%	100.00%

Table 4.1.8 Have you ever shared or used a prompt through any of the following sources? GitHub. Broken out by tenure track

Tenure track	Yes	No
Tenured	2.84%	97.16%
Not tenured but on a tenure track	2.70%	97.30%
Not tenured and not on a tenure track	1.73%	98.27%
Unknown	0.00%	100.00%

Table 4.1.9 Have you ever shared or used a prompt through any of the following sources? GitHub. Broken out by gender of respondent

Gender of respondent	Yes	No
Male	1.47%	98.53%
Female	3.19%	96.81%
Transgender	0.00%	100.00%
Unknown	0.00%	100.00%

Table 4.1.10 Have you ever shared or used a prompt through any of the following sources? GitHub. Broken out by race or ethnicity

Race or ethnicity	Yes	No
White or Caucasian	2.24%	97.76%
Black or African American	3.85%	96.15%
Hispanic or Latino	0.00%	100.00%
Asian or Asian American	5.26%	94.74%
Mixed Race	0.00%	100.00%
Unknown	6.67%	93.33%

Table 4.1.11 Have you ever shared or used a prompt through any of the following sources? GitHub. Broken out by academic field

academic field	Yes	No
Social Sciences (except economics and psychology)	0.00%	100.00%
Education or Library Science	7.14%	92.86%
Allied Health: Nursing, Occupation or Physical Therapy, Nutrition, etc.	0.00%	100.00%
Earth and Environmental Sciences	9.09%	90.91%
Architecture, Fine and Visual Arts	0.00%	100.00%
History	0.00%	100.00%
English, Literature & Languages	2.04%	97.96%
Communications & Journalism	0.00%	100.00%
Engineering	0.00%	100.00%
Philosophy & Religion	0.00%	100.00%
Economics/Finance	0.00%	100.00%
Business (other than economics/finance)	0.00%	100.00%
Performing Arts	5.00%	95.00%
Psychology & Social Work	0.00%	100.00%
Medicine	1.28%	98.72%
Law	0.00%	100.00%
Mathematics/Statistics/Computer Science	8.57%	91.43%
Hard Sciences (Chemistry, Biology, Physics)	4.85%	95.15%
Other	0.00%	100.00%

Table 4.1.12 Have you ever shared or used a prompt through any of the following sources? GitHub. Broken out by years at present institution

Years at present institution	Yes	No
3 years or less	2.50%	97.50%
3+ years to 5 years	7.14%	92.86%
5+ years to 10 years	3.90%	96.10%
10+ years to 20 years	2.49%	97.51%
More than 20 years	1.59%	98.41%

Table 4.2.1 Have you ever shared or used a prompt through any of the following sources? Academic Library

	No Answer	Yes	No
Entire sample	0.00%	3.33%	96.67%

Table 4.2.2 Have you ever shared or used a prompt through any of the following sources? Academic Library. Broken out by enrollment

Enrollment	Yes	No
Less than 5,000	4.76%	95.24%
5,000+ - 17,000	2.40%	97.60%
17,000+ - 32,000	1.19%	98.81%
More than 32,000	5.10%	94.90%

Table 4.2.3 Have you ever shared or used a prompt through any of the following sources? Academic Library. Broken out by type of college or Carnegie Class

Type of college or Carnegie Class	Yes	No
Community College	2.70%	97.30%
BA-Granting Institution	3.64%	96.36%
Specialized College (i.e. performing arts, seminary, etc.)	5.56%	94.44%
MA-PHD Granting College	6.02%	93.98%
Level 1 or Level 2 Research University	2.40%	97.60%

Table 4.2.4 Have you ever shared or used a prompt through any of the following sources? Academic Library. Broken out by public or private college

Public or private college	Yes	No
Public	2.86%	97.14%
Private	4.37%	95.63%

Table 4.2.5 Have you ever shared or used a prompt through any of the following sources? Academic Library. Broken out by age of respondent

age of respondent	Yes	No
30 or under	0.00%	100.00%
31-39	0.00%	100.00%
40-49	2.91%	97.09%
50-59	1.94%	98.06%
60 or over	4.61%	95.39%
Unknown	0.00%	100.00%

Table 4.2.6 Have you ever shared or used a prompt through any of the following sources? Academic Library. Broken out by political views

Political views	Yes	No
Left of Liberal	0.00%	100.00%
Liberal	3.92%	96.08%
Middle of the Road	3.63%	96.37%
Conservative	5.00%	95.00%
Unknown	0.00%	100.00%

Table 4.2.7 Have you ever shared or used a prompt through any of the following sources? Academic Library. Broken out by annual personal income

annual personal income	Yes	No
Less than $50,000	6.25%	93.75%
$50,000 to $75,000	9.23%	90.77%
$75,000+ to $100,000	2.59%	97.41%
$100,000 to $150,000	2.49%	97.51%
More than $150,000	2.83%	97.17%
Unknown	0.00%	100.00%

Table 4.2.8 Have you ever shared or used a prompt through any of the following sources? Academic Library. Broken out by tenure track

Tenure track	Yes	No
Tenured	2.58%	97.42%
Not tenured but on a tenure track	0.00%	100.00%
Not tenured and not on a tenure track	5.19%	94.81%
Unknown	0.00%	100.00%

Table 4.2.9 Have you ever shared or used a prompt through any of the following sources? Academic Library. Broken out by gender of respondent

Gender of respondent	Yes	No
Male	4.78%	95.22%
Female	2.13%	97.87%
Transgender	0.00%	100.00%
Unknown	14.29%	85.71%

Table 4.2.10 Have you ever shared or used a prompt through any of the following sources? Academic Library. Broken out by race or ethnicity

Race or ethnicity	Yes	No
White or Caucasian	3.17%	96.83%
Black or African American	3.85%	96.15%
Hispanic or Latino	0.00%	100.00%
Asian or Asian American	5.26%	94.74%
Mixed Race	0.00%	100.00%
Unknown	13.33%	86.67%

Table 4.2.11 Have you ever shared or used a prompt through any of the following sources? Academic Library. Broken out by academic field

academic field	Yes	No
Social Sciences (except economics and psychology)	1.54%	98.46%
Education or Library Science	11.90%	88.10%
Allied Health: Nursing, Occupation or Physical Therapy, Nutrition, etc.	6.82%	93.18%
Earth and Environmental Sciences	0.00%	100.00%
Architecture, Fine and Visual Arts	10.00%	90.00%
History	7.14%	92.86%
English, Literature & Languages	4.08%	95.92%
Communications & Journalism	9.09%	90.91%
Engineering	0.00%	100.00%
Philosophy & Religion	0.00%	100.00%
Economics/Finance	0.00%	100.00%
Business (other than economics/finance)	6.15%	93.85%
Performing Arts	5.00%	95.00%
Psychology & Social Work	0.00%	100.00%
Medicine	3.85%	96.15%
Law	0.00%	100.00%
Mathematics/Statistics/Computer Science	0.00%	100.00%
Hard Sciences (Chemistry, Biology, Physics)	0.00%	100.00%
Other	0.00%	100.00%

Table 4.2.12 Have you ever shared or used a prompt through any of the following sources? Academic Library. Broken out by years at present institution

Years at present institution	Yes	No
3 years or less	0.00%	100.00%
3+ years to 5 years	10.71%	89.29%
5+ years to 10 years	6.49%	93.51%
10+ years to 20 years	3.48%	96.52%
More than 20 years	2.23%	97.77%

Table 4.3.1 Have you ever shared or used a prompt through any of the following sources? Academic Department.

	No Answer	Yes	No
Entire sample	0.00%	5.76%	94.24%

Table 4.3.2 Have you ever shared or used a prompt through any of the following sources? Academic Department. Broken out by enrollment

Enrollment	Yes	No
Less than 5,000	5.95%	94.05%
5,000+ - 17,000	7.78%	92.22%
17,000+ - 32,000	5.36%	94.64%
More than 32,000	3.82%	96.18%

Table 4.3.3 Have you ever shared or used a prompt through any of the following sources? Academic Department. Broken out by type of college or Carnegie Class

Type of college or Carnegie Class	Yes	No
Community College	5.41%	94.59%
BA-Granting Institution	5.45%	94.55%
Specialized College (i.e. performing arts, seminary, etc.)	5.56%	94.44%
MA-PHD Granting College	7.52%	92.48%
Level 1 or Level 2 Research University	5.28%	94.72%

Table 4.3.4 Have you ever shared or used a prompt through any of the following sources? Academic Department. Broken out by public or private college

Public or private college	Yes	No
Public	5.95%	94.05%
Private	5.34%	94.66%

Table 4.3.5 Have you ever shared or used a prompt through any of the following sources? Academic Department. Broken out by age of respondent

age of respondent	Yes	No
30 or under	0.00%	100.00%
31-39	5.00%	95.00%
40-49	4.85%	95.15%
50-59	5.16%	94.84%
60 or over	6.63%	93.37%
Unknown	0.00%	100.00%

Table 4.3.6 Have you ever shared or used a prompt through any of the following sources? Academic Department. Broken out by political views

Political views	Yes	No
Left of Liberal	2.97%	97.03%
Liberal	4.31%	95.69%
Middle of the Road	9.33%	90.67%
Conservative	6.00%	94.00%
Unknown	0.00%	100.00%

Table 4.3.7 Have you ever shared or used a prompt through any of the following sources? Academic Department. Broken out by annual personal income

annual personal income	Yes	No
Less than $50,000	6.25%	93.75%
$50,000 to $75,000	9.23%	90.77%
$75,000+ to $100,000	6.90%	93.10%
$100,000 to $150,000	5.47%	94.53%
More than $150,000	4.86%	95.14%
Unknown	0.00%	100.00%

Table 4.3.8 Have you ever shared or used a prompt through any of the following sources? Academic Department. Broken out by tenure track

Tenure track	Yes	No
Tenured	5.15%	94.85%
Not tenured but on a tenure track	2.70%	97.30%
Not tenured and not on a tenure track	7.36%	92.64%
Unknown	0.00%	100.00%

Table 4.3.9 Have you ever shared or used a prompt through any of the following sources? Academic Department. Broken out by gender of respondent

Gender of respondent	Yes	No
Male	5.88%	94.12%
Female	5.85%	94.15%
Transgender	0.00%	100.00%
Unknown	0.00%	100.00%

Table 4.3.10 Have you ever shared or used a prompt through any of the following sources? Academic Department. Broken out by race or ethnicity

Race or ethnicity	Yes	No
White or Caucasian	5.60%	94.40%
Black or African American	11.54%	88.46%
Hispanic or Latino	3.70%	96.30%
Asian or Asian American	5.26%	94.74%
Mixed Race	0.00%	100.00%
Unknown	13.33%	86.67%

Table 4.3.11 Have you ever shared or used a prompt through any of the following sources? Academic Department. Broken out by academic field

academic field	Yes	No
Social Sciences (except economics and psychology)	3.08%	96.92%
Education or Library Science	4.76%	95.24%
Allied Health: Nursing, Occupation or Physical Therapy, Nutrition, etc.	2.27%	97.73%
Earth and Environmental Sciences	4.55%	95.45%
Architecture, Fine and Visual Arts	0.00%	100.00%
History	14.29%	85.71%
English, Literature & Languages	6.12%	93.88%
Communications & Journalism	0.00%	100.00%
Engineering	8.11%	91.89%
Philosophy & Religion	0.00%	100.00%
Economics/Finance	5.56%	94.44%
Business (other than economics/finance)	15.38%	84.62%
Performing Arts	5.00%	95.00%
Psychology & Social Work	9.68%	90.32%
Medicine	5.13%	94.87%
Law	0.00%	100.00%
Mathematics/Statistics/Computer Science	5.71%	94.29%
Hard Sciences (Chemistry, Biology, Physics)	1.94%	98.06%
Other	25.00%	75.00%

Table 4.3.12 Have you ever shared or used a prompt through any of the following sources? Academic Department. Broken out by years at present institution

Years at present institution	Yes	No
3 years or less	2.50%	97.50%
3+ years to 5 years	3.57%	96.43%
5+ years to 10 years	5.19%	94.81%
10+ years to 20 years	5.47%	94.53%
More than 20 years	6.69%	93.31%

Table 4.4.1 Have you ever shared or used a prompt through any of the following sources? Institutional Digital Repository.

	No Answer	Yes	No
Entire sample	0.00%	1.52%	98.48%

Table 4.4.2 Have you ever shared or used a prompt through any of the following sources? Institutional Digital Repository. Broken out by enrollment

Enrollment	Yes	No
Less than 5,000	0.60%	99.40%
5,000+ - 17,000	2.40%	97.60%
17,000+ - 32,000	1.79%	98.21%
More than 32,000	1.27%	98.73%

Table 4.4.3 Have you ever shared or used a prompt through any of the following sources? Institutional Digital Repository. Broken out by type of college or Carnegie Class

Type of college or Carnegie Class	Yes	No
Community College	2.70%	97.30%
BA-Granting Institution	0.00%	100.00%
Specialized College (i.e. performing arts, seminary, etc.)	0.00%	100.00%
MA-PHD Granting College	3.01%	96.99%
Level 1 or Level 2 Research University	1.20%	98.80%

Table 4.4.4 Have you ever shared or used a prompt through any of the following sources? Institutional Digital Repository. Broken out by public or private college

Public or private college	Yes	No
Public	1.54%	98.46%
Private	1.46%	98.54%

Table 4.4.5 Have you ever shared or used a prompt through any of the following sources? Institutional Digital Repository. Broken out by age of respondent

age of respondent	Yes	No
30 or under	0.00%	100.00%
31-39	2.50%	97.50%
40-49	0.00%	100.00%
50-59	1.94%	98.06%
60 or over	1.73%	98.27%
Unknown	0.00%	100.00%

Table 4.4.6 Have you ever shared or used a prompt through any of the following sources? Institutional Digital Repository. Broken out by political views

Political views	Yes	No
Left of Liberal	0.00%	100.00%
Liberal	1.96%	98.04%
Middle of the Road	1.04%	98.96%
Conservative	2.00%	98.00%
Unknown	9.09%	90.91%

Table 4.4.7 Have you ever shared or used a prompt through any of the following sources? Institutional Digital Repository. Broken out by annual personal income

annual personal income	Yes	No
Less than $50,000	6.25%	93.75%
$50,000 to $75,000	3.08%	96.92%
$75,000+ to $100,000	0.86%	99.14%
$100,000 to $150,000	1.49%	98.51%
More than $150,000	1.21%	98.79%
Unknown	0.00%	100.00%

Table 4.4.8 Have you ever shared or used a prompt through any of the following sources? Institutional Digital Repository. Broken out by tenure track

Tenure track	Yes	No
Tenured	1.03%	98.97%
Not tenured but on a tenure track	2.70%	97.30%
Not tenured and not on a tenure track	2.16%	97.84%
Unknown	0.00%	100.00%

Table 4.4.9 Have you ever shared or used a prompt through any of the following sources? Institutional Digital Repository. Broken out by gender of respondent

Gender of respondent	Yes	No
Male	2.21%	97.79%
Female	1.06%	98.94%
Transgender	0.00%	100.00%
Unknown	0.00%	100.00%

Table 4.4.10 Have you ever shared or used a prompt through any of the following sources? Institutional Digital Repository. Broken out by race or ethnicity

Race or ethnicity	Yes	No
White or Caucasian	1.49%	98.51%
Black or African American	0.00%	100.00%
Hispanic or Latino	0.00%	100.00%
Asian or Asian American	5.26%	94.74%
Mixed Race	0.00%	100.00%
Unknown	0.00%	100.00%

Table 4.4.11 Have you ever shared or used a prompt through any of the following sources? Institutional Digital Repository. Broken out by academic field

academic field	Yes	No
Social Sciences (except economics and psychology)	0.00%	100.00%
Education or Library Science	7.14%	92.86%
Allied Health: Nursing, Occupation or Physical Therapy, Nutrition, etc.	0.00%	100.00%
Earth and Environmental Sciences	0.00%	100.00%
Architecture, Fine and Visual Arts	0.00%	100.00%
History	0.00%	100.00%
English, Literature & Languages	0.00%	100.00%
Communications & Journalism	0.00%	100.00%
Engineering	2.70%	97.30%
Philosophy & Religion	0.00%	100.00%
Economics/Finance	0.00%	100.00%
Business (other than economics/finance)	3.08%	96.92%
Performing Arts	5.00%	95.00%
Psychology & Social Work	0.00%	100.00%
Medicine	0.00%	100.00%
Law	0.00%	100.00%
Mathematics/Statistics/Computer Science	8.57%	91.43%
Hard Sciences (Chemistry, Biology, Physics)	0.00%	100.00%
Other	0.00%	100.00%

Table 4.4.12 Have you ever shared or used a prompt through any of the following sources? Institutional Digital Repository. Broken out by years at present institution

Years at present institution	Yes	No
3 years or less	2.50%	97.50%
3+ years to 5 years	0.00%	100.00%
5+ years to 10 years	1.30%	98.70%
10+ years to 20 years	1.99%	98.01%
More than 20 years	1.27%	98.73%

Table 4.5.1 Have you ever shared or used a prompt through any of the following sources? Instructional Design or Teaching and Learning Center.

	No Answer	Yes	No
Entire sample	0.00%	4.09%	95.91%

Table 4.5.2 Have you ever shared or used a prompt through any of the following sources? Instructional Design or Teaching and Learning Center. Broken out by enrollment

Enrollment	Yes	No
Less than 5,000	5.95%	94.05%
5,000+ - 17,000	4.19%	95.81%
17,000+ - 32,000	2.38%	97.62%
More than 32,000	3.82%	96.18%

Table 4.5.3 Have you ever shared or used a prompt through any of the following sources? Instructional Design or Teaching and Learning Center. Broken out by type of college or Carnegie Class

Type of college or Carnegie Class	Yes	No
Community College	2.70%	97.30%
BA-Granting Institution	1.82%	98.18%
Specialized College (i.e. performing arts, seminary, etc.)	5.56%	94.44%
MA-PHD Granting College	5.26%	94.74%
Level 1 or Level 2 Research University	4.08%	95.92%

Table 4.5.4 Have you ever shared or used a prompt through any of the following sources? Instructional Design or Teaching and Learning Center. Broken out by public or private college

Public or private college	Yes	No
Public	3.74%	96.26%
Private	4.85%	95.15%

Table 4.5.5 Have you ever shared or used a prompt through any of the following sources? Instructional Design or Teaching and Learning Center. Broken out by age of respondent

age of respondent	Yes	No
30 or under	0.00%	100.00%
31-39	7.50%	92.50%
40-49	0.97%	99.03%
50-59	1.94%	98.06%
60 or over	5.48%	94.52%
Unknown	16.67%	83.33%

Table 4.5.6 Have you ever shared or used a prompt through any of the following sources? Instructional Design or Teaching and Learning Center. Broken out by political views

Political views	Yes	No
Left of Liberal	3.96%	96.04%
Liberal	4.31%	95.69%
Middle of the Road	2.59%	97.41%
Conservative	6.00%	94.00%
Unknown	9.09%	90.91%

Table 4.5.7 Have you ever shared or used a prompt through any of the following sources? Instructional Design or Teaching and Learning Center. Broken out by annual personal income

annual personal income	Yes	No
Less than $50,000	6.25%	93.75%
$50,000 to $75,000	4.62%	95.38%
$75,000+ to $100,000	2.59%	97.41%
$100,000 to $150,000	4.48%	95.52%
More than $150,000	3.64%	96.36%
Unknown	13.33%	86.67%

Table 4.5.8 Have you ever shared or used a prompt through any of the following sources? Instructional Design or Teaching and Learning Center. Broken out by tenure track

Tenure track	Yes	No
Tenured	3.35%	96.65%
Not tenured but on a tenure track	2.70%	97.30%
Not tenured and not on a tenure track	5.19%	94.81%
Unknown	25.00%	75.00%

Table 4.5.9 Have you ever shared or used a prompt through any of the following sources? Instructional Design or Teaching and Learning Center. Broken out by gender of respondent

Gender of respondent	Yes	No
Male	6.99%	93.01%
Female	1.86%	98.14%
Transgender	0.00%	100.00%
Unknown	14.29%	85.71%

Table 4.5.10 Have you ever shared or used a prompt through any of the following sources? Instructional Design or Teaching and Learning Center. Broken out by race or ethnicity

Race or ethnicity	Yes	No
White or Caucasian	4.10%	95.90%
Black or African American	7.69%	92.31%
Hispanic or Latino	3.70%	96.30%
Asian or Asian American	2.63%	97.37%
Mixed Race	0.00%	100.00%
Unknown	6.67%	93.33%

Table 4.5.11 Have you ever shared or used a prompt through any of the following sources? Instructional Design or Teaching and Learning Center. Broken out by academic field

academic field	Yes	No
Social Sciences (except economics and psychology)	0.00%	100.00%
Education or Library Science	11.90%	88.10%
Allied Health: Nursing, Occupation or Physical Therapy, Nutrition, etc.	11.36%	88.64%
Earth and Environmental Sciences	0.00%	100.00%
Architecture, Fine and Visual Arts	0.00%	100.00%
History	0.00%	100.00%
English, Literature & Languages	2.04%	97.96%
Communications & Journalism	0.00%	100.00%
Engineering	5.41%	94.59%
Philosophy & Religion	0.00%	100.00%
Economics/Finance	0.00%	100.00%
Business (other than economics/finance)	6.15%	93.85%
Performing Arts	10.00%	90.00%
Psychology & Social Work	0.00%	100.00%
Medicine	5.13%	94.87%
Law	0.00%	100.00%
Mathematics/Statistics/Computer Science	2.86%	97.14%
Hard Sciences (Chemistry, Biology, Physics)	2.91%	97.09%
Other	0.00%	100.00%

Table 4.5.12 Have you ever shared or used a prompt through any of the following sources? Instructional Design or Teaching and Learning Center. Broken out by years at present institution

Years at present institution	Yes	No
3 years or less	2.50%	97.50%
3+ years to 5 years	7.14%	92.86%
5+ years to 10 years	2.60%	97.40%
10+ years to 20 years	4.48%	95.52%
More than 20 years	4.14%	95.86%

Table 4.6.1 Have you ever shared or used a prompt through any of the following sources? Professional or Trade Association.

	No Answer	Yes	No
Entire sample	0.00%	3.64%	96.36%

Table 4.6.2 Have you ever shared or used a prompt through any of the following sources? Professional or Trade Association. Broken out by enrollment

Enrollment	Yes	No
Less than 5,000	5.95%	94.05%
5,000+ - 17,000	1.80%	98.20%
17,000+ - 32,000	2.38%	97.62%
More than 32,000	4.46%	95.54%

Table 4.6.3 Have you ever shared or used a prompt through any of the following sources? Professional or Trade Association. Broken out by type of college or Carnegie Class

Type of college or Carnegie Class	Yes	No
Community College	0.00%	100.00%
BA-Granting Institution	5.45%	94.55%
Specialized College (i.e. performing arts, seminary, etc.)	5.56%	94.44%
MA-PHD Granting College	4.51%	95.49%
Level 1 or Level 2 Research University	3.36%	96.64%

Table 4.6.4 Have you ever shared or used a prompt through any of the following sources? Professional or Trade Association. Broken out by public or private college

Public or private college	Yes	No
Public	3.30%	96.70%
Private	4.37%	95.63%

Table 4.6.5 Have you ever shared or used a prompt through any of the following sources? Professional or Trade Association. Broken out by age of respondent

age of respondent	Yes	No
30 or under	0.00%	100.00%
31-39	0.00%	100.00%
40-49	2.91%	97.09%
50-59	1.94%	98.06%
60 or over	5.19%	94.81%
Unknown	0.00%	100.00%

Table 4.6.6 Have you ever shared or used a prompt through any of the following sources? Professional or Trade Association. Broken out by political views

Political views	Yes	No
Left of Liberal	1.98%	98.02%
Liberal	2.75%	97.25%
Middle of the Road	5.70%	94.30%
Conservative	4.00%	96.00%
Unknown	0.00%	100.00%

Table 4.6.7 Have you ever shared or used a prompt through any of the following sources? Professional or Trade Association. Broken out by annual personal income

annual personal income	Yes	No
Less than $50,000	0.00%	100.00%
$50,000 to $75,000	6.15%	93.85%
$75,000+ to $100,000	6.90%	93.10%
$100,000 to $150,000	2.49%	97.51%
More than $150,000	2.83%	97.17%
Unknown	0.00%	100.00%

Table 4.6.8 Have you ever shared or used a prompt through any of the following sources? Professional or Trade Association. Broken out by tenure track

Tenure track	Yes	No
Tenured	2.32%	97.68%
Not tenured but on a tenure track	2.70%	97.30%
Not tenured and not on a tenure track	6.06%	93.94%
Unknown	0.00%	100.00%

Table 4.6.9 Have you ever shared or used a prompt through any of the following sources? Professional or Trade Association. Broken out by gender of respondent

Gender of respondent	Yes	No
Male	4.04%	95.96%
Female	3.19%	96.81%
Transgender	20.00%	80.00%
Unknown	0.00%	100.00%

Table 4.6.10 Have you ever shared or used a prompt through any of the following sources? Professional or Trade Association. Broken out by race or ethnicity

Race or ethnicity	Yes	No
White or Caucasian	3.73%	96.27%
Black or African American	0.00%	100.00%
Hispanic or Latino	0.00%	100.00%
Asian or Asian American	7.89%	92.11%
Mixed Race	5.56%	94.44%
Unknown	0.00%	100.00%

Table 4.6.11 Have you ever shared or used a prompt through any of the following sources? Professional or Trade Association. Broken out by academic field

academic field	Yes	No
Social Sciences (except economics and psychology)	4.62%	95.38%
Education or Library Science	14.29%	85.71%
Allied Health: Nursing, Occupation or Physical Therapy, Nutrition, etc.	2.27%	97.73%
Earth and Environmental Sciences	4.55%	95.45%
Architecture, Fine and Visual Arts	0.00%	100.00%
History	7.14%	92.86%
English, Literature & Languages	2.04%	97.96%
Communications & Journalism	18.18%	81.82%
Engineering	0.00%	100.00%
Philosophy & Religion	0.00%	100.00%
Economics/Finance	5.56%	94.44%
Business (other than economics/finance)	7.69%	92.31%
Performing Arts	5.00%	95.00%
Psychology & Social Work	0.00%	100.00%
Medicine	0.00%	100.00%
Law	0.00%	100.00%
Mathematics/Statistics/Computer Science	0.00%	100.00%
Hard Sciences (Chemistry, Biology, Physics)	1.94%	98.06%
Other	0.00%	100.00%

Table 4.6.12 Have you ever shared or used a prompt through any of the following sources? Professional or Trade Association. Broken out by years at present institution

Years at present institution	Yes	No
3 years or less	2.50%	97.50%
3+ years to 5 years	3.57%	96.43%
5+ years to 10 years	5.19%	94.81%
10+ years to 20 years	4.98%	95.02%
More than 20 years	2.55%	97.45%

Table 4.7.1 Have you ever shared or used a prompt through any of the following sources? Other (Describe).

	No Answer	Yes	No
Entire sample	0.00%	4.09%	95.91%

Table 4.7.2 Have you ever shared or used a prompt through any of the following sources? Other (Describe). Broken out by enrollment

Enrollment	Yes	No
Less than 5,000	4.76%	95.24%
5,000+ - 17,000	3.59%	96.41%
17,000+ - 32,000	4.76%	95.24%
More than 32,000	3.18%	96.82%

Table 4.7.3 Have you ever shared or used a prompt through any of the following sources? Other (Describe). Broken out by type of college or Carnegie Class

Type of college or Carnegie Class	Yes	No
Community College	5.41%	94.59%
BA-Granting Institution	5.45%	94.55%
Specialized College (i.e. performing arts, seminary, etc.)	0.00%	100.00%
MA-PHD Granting College	6.02%	93.98%
Level 1 or Level 2 Research University	3.36%	96.64%

Table 4.7.4 Have you ever shared or used a prompt through any of the following sources? Other (Describe). Broken out by public or private college

Public or private college	Yes	No
Public	4.19%	95.81%
Private	3.88%	96.12%

Table 4.7.5 Have you ever shared or used a prompt through any of the following sources? Other (Describe). Broken out by age of respondent

age of respondent	Yes	No
30 or under	0.00%	100.00%
31-39	5.00%	95.00%
40-49	10.68%	89.32%
50-59	2.58%	97.42%
60 or over	2.88%	97.12%
Unknown	0.00%	100.00%

Table 4.7.6 Have you ever shared or used a prompt through any of the following sources? Other (Describe). Broken out by political views

Political views	Yes	No
Left of Liberal	1.98%	98.02%
Liberal	3.92%	96.08%
Middle of the Road	5.18%	94.82%
Conservative	5.00%	95.00%
Unknown	0.00%	100.00%

Table 4.7.7 Have you ever shared or used a prompt through any of the following sources? Other (Describe). Broken out by annual personal income

annual personal income	Yes	No
Less than $50,000	0.00%	100.00%
$50,000 to $75,000	6.15%	93.85%
$75,000+ to $100,000	0.86%	99.14%
$100,000 to $150,000	4.48%	95.52%
More than $150,000	5.26%	94.74%
Unknown	0.00%	100.00%

Table 4.7.8 Have you ever shared or used a prompt through any of the following sources? Other (Describe). Broken out by tenure track

Tenure track	Yes	No
Tenured	3.09%	96.91%
Not tenured but on a tenure track	2.70%	97.30%
Not tenured and not on a tenure track	6.06%	93.94%
Unknown	0.00%	100.00%

Table 4.7.9 Have you ever shared or used a prompt through any of the following sources? Other (Describe). Broken out by gender of respondent

Gender of respondent	Yes	No
Male	2.57%	97.43%
Female	5.32%	94.68%
Transgender	0.00%	100.00%
Unknown	0.00%	100.00%

Table 4.7.10 Have you ever shared or used a prompt through any of the following sources? Other (Describe). Broken out by race or ethnicity

Race or ethnicity	Yes	No
White or Caucasian	3.92%	96.08%
Black or African American	0.00%	100.00%
Hispanic or Latino	3.70%	96.30%
Asian or Asian American	10.53%	89.47%
Mixed Race	5.56%	94.44%
Unknown	0.00%	100.00%

Table 4.7.11 Have you ever shared or used a prompt through any of the following sources? Other (Describe). Broken out by academic field

academic field	Yes	No
Social Sciences (except economics and psychology)	3.08%	96.92%
Education or Library Science	4.76%	95.24%
Allied Health: Nursing, Occupation or Physical Therapy, Nutrition, etc.	2.27%	97.73%
Earth and Environmental Sciences	0.00%	100.00%
Architecture, Fine and Visual Arts	0.00%	100.00%
History	0.00%	100.00%
English, Literature & Languages	4.08%	95.92%
Communications & Journalism	0.00%	100.00%
Engineering	8.11%	91.89%
Philosophy & Religion	0.00%	100.00%
Economics/Finance	5.56%	94.44%
Business (other than economics/finance)	7.69%	92.31%
Performing Arts	5.00%	95.00%
Psychology & Social Work	3.23%	96.77%
Medicine	8.97%	91.03%
Law	0.00%	100.00%
Mathematics/Statistics/Computer Science	0.00%	100.00%
Hard Sciences (Chemistry, Biology, Physics)	1.94%	98.06%
Other	0.00%	100.00%

Table 4.7.12 Have you ever shared or used a prompt through any of the following sources? Other (Describe). Broken out by years at present institution

Years at present institution	Yes	No
3 years or less	7.50%	92.50%
3+ years to 5 years	14.29%	85.71%
5+ years to 10 years	1.30%	98.70%
10+ years to 20 years	4.98%	95.02%
More than 20 years	2.87%	97.13%

Table 5.1 Have you ever included details of prompt engineering in any academic paper or article you have written?

	No Answer	Yes	No
Entire sample	5.61%	2.73%	91.67%

Table 5.2 Have you ever included details of prompt engineering in any academic paper or article you have written? Broken out by enrollment

Enrollment	No Answer	Yes	No
Less than 5,000	4.17%	2.38%	93.45%
5,000+ - 17,000	7.19%	2.40%	90.42%
17,000+ - 32,000	5.36%	2.98%	91.67%
More than 32,000	5.73%	3.18%	91.08%

Table 5.3 Have you ever included details of prompt engineering in any academic paper or article you have written? Broken out by type of college or Carnegie Class

Type of college or Carnegie Class	No Answer	Yes	No
Community College	10.81%	2.70%	86.49%
BA-Granting Institution	3.64%	0.00%	96.36%
Specialized College (i.e. performing arts, seminary, etc.)	0.00%	5.56%	94.44%
MA-PHD Granting College	6.77%	3.76%	89.47%
Level 1 or Level 2 Research University	5.28%	2.64%	92.09%

Table 5.4 Have you ever included details of prompt engineering in any academic paper or article you have written? Broken out by public or private college

Public or private college	No Answer	Yes	No
Public	6.61%	2.64%	90.75%
Private	3.40%	2.91%	93.69%

Table 5.5 Have you ever included details of prompt engineering in any academic paper or article you have written? Broken out by age of respondent

age of respondent	No Answer	Yes	No
30 or under	0.00%	0.00%	100.00%
31-39	5.00%	0.00%	95.00%
40-49	5.83%	1.94%	92.23%
50-59	8.39%	3.23%	88.39%
60 or over	4.61%	2.88%	92.51%
Unknown	0.00%	16.67%	83.33%

Table 5.6 Have you ever included details of prompt engineering in any academic paper or article you have written? Broken out by political views

Political views	No Answer	Yes	No
Left of Liberal	5.94%	1.98%	92.08%
Liberal	4.71%	3.53%	91.76%
Middle of the Road	7.77%	2.59%	89.64%
Conservative	4.00%	1.00%	95.00%
Unknown	0.00%	9.09%	90.91%

Table 5.7 Have you ever included details of prompt engineering in any academic paper or article you have written? Broken out by annual personal income

annual personal income	No Answer	Yes	No
Less than $50,000	0.00%	6.25%	93.75%
$50,000 to $75,000	1.54%	6.15%	92.31%
$75,000+ to $100,000	6.90%	0.86%	92.24%
$100,000 to $150,000	6.97%	1.49%	91.54%
More than $150,000	4.86%	3.64%	91.50%
Unknown	13.33%	0.00%	86.67%

Table 5.8 Have you ever included details of prompt engineering in any academic paper or article you have written? Broken out by tenure track

Tenure track	No Answer	Yes	No
Tenured	6.19%	2.84%	90.98%
Not tenured but on a tenure track	2.70%	2.70%	94.59%
Not tenured and not on a tenure track	4.76%	2.60%	92.64%
Unknown	25.00%	0.00%	75.00%

Table 5.9 Have you ever included details of prompt engineering in any academic paper or article you have written? Broken out by gender of respondent

Gender of respondent	No Answer	Yes	No
Male	5.15%	3.31%	91.54%
Female	5.59%	2.39%	92.02%
Transgender	20.00%	0.00%	80.00%
Unknown	14.29%	0.00%	85.71%

Table 5.10 Have you ever included details of prompt engineering in any academic paper or article you have written? Broken out by race or ethnicity

Race or ethnicity	No Answer	Yes	No
White or Caucasian	5.41%	2.24%	92.35%
Black or African American	3.85%	7.69%	88.46%
Hispanic or Latino	0.00%	7.41%	92.59%
Asian or Asian American	10.53%	5.26%	84.21%
Mixed Race	0.00%	0.00%	100.00%
Unknown	20.00%	0.00%	80.00%

Table 5.11 Have you ever included details of prompt engineering in any academic paper or article you have written? Broken out by academic field

academic field	No Answer	Yes	No
Social Sciences (except economics and psychology)	4.62%	4.62%	90.77%
Education or Library Science	2.38%	11.90%	85.71%
Allied Health: Nursing, Occupation or Physical Therapy, Nutrition, etc.	6.82%	0.00%	93.18%
Earth and Environmental Sciences	4.55%	4.55%	90.91%
Architecture, Fine and Visual Arts	0.00%	0.00%	100.00%
History	0.00%	0.00%	100.00%
English, Literature & Languages	6.12%	2.04%	91.84%
Communications & Journalism	0.00%	0.00%	100.00%
Engineering	16.22%	0.00%	83.78%
Philosophy & Religion	0.00%	0.00%	100.00%
Economics/Finance	0.00%	0.00%	100.00%
Business (other than economics/finance)	6.15%	4.62%	89.23%
Performing Arts	0.00%	0.00%	100.00%
Psychology & Social Work	9.68%	6.45%	83.87%
Medicine	6.41%	1.28%	92.31%
Law	0.00%	0.00%	100.00%
Mathematics/Statistics/Computer Science	8.57%	2.86%	88.57%
Hard Sciences (Chemistry, Biology, Physics)	4.85%	0.97%	94.17%
Other	0.00%	0.00%	100.00%

Table 5.12 Have you ever included details of prompt engineering in any academic paper or article you have written? Broken out by years at present institution

Years at present institution	No Answer	Yes	No
3 years or less	5.00%	5.00%	90.00%
3+ years to 5 years	3.57%	0.00%	96.43%
5+ years to 10 years	1.30%	2.60%	96.10%
10+ years to 20 years	6.47%	2.99%	90.55%
More than 20 years	6.37%	2.55%	91.08%

Table 6.1 How satisfied are you with the availability of libraries or archives of artificial intelligence prompts?

	No Answer	Very satisfied	Satisfied	Neither satisfied nor dissatisfied	Dissatisfied	Very dissatisfied
Entire sample	9.70%	1.52%	6.36%	76.97%	3.64%	1.82%

Table 6.2 How satisfied are you with the availability of libraries or archives of artificial intelligence prompts? Broken out by enrollment

Enrollment	No Answer	Very satisfied	Satisfied	Neither satisfied nor dissatisfied	Dissatisfied	Very dissatisfied
Less than 5,000	13.10%	1.79%	8.93%	71.43%	2.98%	1.79%
5,000+ - 17,000	9.58%	0.60%	6.59%	78.44%	2.99%	1.80%
17,000+ - 32,000	8.33%	1.79%	4.17%	80.95%	2.98%	1.79%
More than 32,000	7.64%	1.91%	5.73%	77.07%	5.73%	1.91%

Table 6.3 How satisfied are you with the availability of libraries or archives of artificial intelligence prompts? Broken out by type of college or Carnegie Class

Type of college or Carnegie Class	No Answer	Very satisfied	Satisfied	Neither satisfied nor dissatisfied	Dissatisfied	Very dissatisfied
Community College	8.11%	0.00%	5.41%	81.08%	5.41%	0.00%
BA-Granting Institution	10.91%	0.00%	5.45%	69.09%	10.91%	3.64%
Specialized College (i.e. performing arts, seminary, etc.)	11.11%	0.00%	27.78%	61.11%	0.00%	0.00%
MA-PHD Granting College	10.53%	2.26%	6.77%	73.68%	3.01%	3.76%
Level 1 or Level 2 Research University	9.35%	1.68%	5.52%	79.38%	2.88%	1.20%

Table 6.4 How satisfied are you with the availability of libraries or archives of artificial intelligence prompts? Broken out by public or private college

Public or private college	No Answer	Very satisfied	Satisfied	Neither satisfied nor dissatisfied	Dissatisfied	Very dissatisfied
Public	8.59%	1.98%	5.07%	78.41%	3.96%	1.98%
Private	12.14%	0.49%	9.22%	73.79%	2.91%	1.46%

Table 6.5 How satisfied are you with the availability of libraries or archives of artificial intelligence prompts? Broken out by age of respondent

age of respondent	No Answer	Very satisfied	Satisfied	Neither satisfied nor dissatisfied	Dissatisfied	Very dissatisfied
30 or under	11.11%	0.00%	0.00%	77.78%	11.11%	0.00%
31-39	2.50%	0.00%	2.50%	85.00%	7.50%	2.50%
40-49	10.68%	0.00%	3.88%	78.64%	5.83%	0.97%
50-59	11.61%	0.65%	6.45%	76.13%	2.58%	2.58%
60 or over	9.22%	2.59%	7.49%	76.08%	2.88%	1.73%
Unknown	16.67%	0.00%	16.67%	66.67%	0.00%	0.00%

Table 6.6 How satisfied are you with the availability of libraries or archives of artificial intelligence prompts? Broken out by political views

Political views	No Answer	Very satisfied	Satisfied	Neither satisfied nor dissatisfied	Dissatisfied	Very dissatisfied
Left of Liberal	9.90%	0.00%	4.95%	78.22%	2.97%	3.96%
Liberal	9.41%	1.57%	9.02%	76.47%	3.14%	0.39%
Middle of the Road	9.33%	0.52%	4.66%	78.24%	4.15%	3.11%
Conservative	11.00%	4.00%	5.00%	75.00%	4.00%	1.00%
Unknown	9.09%	9.09%	0.00%	72.73%	9.09%	0.00%

Table 6.7 How satisfied are you with the availability of libraries or archives of artificial intelligence prompts? Broken out by annual personal income

annual personal income	No Answer	Very satisfied	Satisfied	Neither satisfied nor dissatisfied	Dissatisfied	Very dissatisfied
Less than $50,000	6.25%	6.25%	6.25%	75.00%	6.25%	0.00%
$50,000 to $75,000	6.15%	0.00%	9.23%	78.46%	4.62%	1.54%
$75,000+ to $100,000	10.34%	2.59%	4.31%	75.00%	2.59%	5.17%
$100,000 to $150,000	10.95%	0.50%	5.47%	80.10%	2.49%	0.50%
More than $150,000	9.31%	1.62%	7.69%	76.52%	3.64%	1.21%
Unknown	13.33%	6.67%	0.00%	53.33%	20.00%	6.67%

Table 6.8 How satisfied are you with the availability of libraries or archives of artificial intelligence prompts? Broken out by tenure track

Tenure track	No Answer	Very satisfied	Satisfied	Neither satisfied nor dissatisfied	Dissatisfied	Very dissatisfied
Tenured	11.34%	1.29%	4.12%	78.87%	2.58%	1.80%
Not tenured but on a tenure track	8.11%	0.00%	5.41%	72.97%	10.81%	2.70%
Not tenured and not on a tenure track	6.93%	2.16%	9.96%	75.32%	3.90%	1.73%
Unknown	25.00%	0.00%	25.00%	25.00%	25.00%	0.00%

Table 6.9 How satisfied are you with the availability of libraries or archives of artificial intelligence prompts? Broken out by gender of respondent

Gender of respondent	No Answer	Very satisfied	Satisfied	Neither satisfied nor dissatisfied	Dissatisfied	Very dissatisfied
Male	8.82%	1.47%	7.72%	74.63%	4.78%	2.57%
Female	10.37%	1.60%	5.59%	78.99%	2.13%	1.33%
Transgender	20.00%	0.00%	0.00%	60.00%	20.00%	0.00%
Unknown	0.00%	0.00%	0.00%	71.43%	28.57%	0.00%

Table 6.10 How satisfied are you with the availability of libraries or archives of artificial intelligence prompts? Broken out by race or ethnicity

Race or ethnicity	No Answer	Very satisfied	Satisfied	Neither satisfied nor dissatisfied	Dissatisfied	Very dissatisfied
White or Caucasian	9.89%	1.49%	6.34%	76.87%	3.54%	1.87%
Black or African American	0.00%	0.00%	11.54%	88.46%	0.00%	0.00%
Hispanic or Latino	3.70%	0.00%	3.70%	81.48%	7.41%	3.70%
Asian or Asian American	23.68%	2.63%	5.26%	63.16%	5.26%	0.00%
Mixed Race	0.00%	0.00%	11.11%	83.33%	0.00%	5.56%
Unknown	6.67%	6.67%	0.00%	80.00%	6.67%	0.00%

Table 6.11 How satisfied are you with the availability of libraries or archives of artificial intelligence prompts? Broken out by academic field

academic field	No Answer	Very satisfied	Satisfied	Neither satisfied nor dissatisfied	Dissatisfied	Very dissatisfied
Social Sciences (except economics and psychology)	12.31%	0.00%	4.62%	73.85%	4.62%	4.62%
Education or Library Science	4.76%	4.76%	9.52%	71.43%	2.38%	7.14%
Allied Health: Nursing, Occupation or Physical Therapy, Nutrition, etc.	4.55%	0.00%	11.36%	81.82%	2.27%	0.00%
Earth and Environmental Sciences	9.09%	4.55%	4.55%	81.82%	0.00%	0.00%
Architecture, Fine and Visual Arts	20.00%	0.00%	20.00%	60.00%	0.00%	0.00%
History	14.29%	0.00%	7.14%	78.57%	0.00%	0.00%
English, Literature & Languages	8.16%	0.00%	4.08%	83.67%	4.08%	0.00%
Communications & Journalism	0.00%	0.00%	9.09%	81.82%	9.09%	0.00%
Engineering	13.51%	2.70%	8.11%	72.97%	2.70%	0.00%
Philosophy & Religion	0.00%	0.00%	0.00%	83.33%	0.00%	16.67%
Economics/Finance	5.56%	0.00%	0.00%	83.33%	11.11%	0.00%
Business (other than economics/finance)	9.23%	3.08%	9.23%	67.69%	9.23%	1.54%
Performing Arts	10.00%	0.00%	5.00%	85.00%	0.00%	0.00%
Psychology & Social Work	12.90%	0.00%	6.45%	70.97%	6.45%	3.23%
Medicine	10.26%	1.28%	6.41%	75.64%	3.85%	2.56%
Law	33.33%	0.00%	16.67%	50.00%	0.00%	0.00%
Mathematics/Statistics/Computer Science	14.29%	5.71%	2.86%	74.29%	0.00%	2.86%
Hard Sciences (Chemistry, Biology, Physics)	7.77%	0.97%	2.91%	86.41%	1.94%	0.00%
Other	25.00%	0.00%	25.00%	50.00%	0.00%	0.00%

Table 6.12 How satisfied are you with the availability of libraries or archives of artificial intelligence prompts? Broken out by years at present institution

Years at present institution	No Answer	Very satisfied	Satisfied	Neither satisfied nor dissatisfied	Dissatisfied	Very dissatisfied
3 years or less	10.00%	0.00%	5.00%	77.50%	5.00%	2.50%
3+ years to 5 years	3.57%	0.00%	3.57%	92.86%	0.00%	0.00%
5+ years to 10 years	5.19%	1.30%	5.19%	83.12%	3.90%	1.30%
10+ years to 20 years	10.45%	1.49%	6.47%	74.13%	4.98%	2.49%
More than 20 years	10.83%	1.91%	7.01%	75.80%	2.87%	1.59%

www.ingramcontent.com/pod-product-compliance
Lightning Source LLC
LaVergne TN
LVHW082358200926
848481LV00012B/3606